All That Lies Between

Austin P. Torney

Copyright

© 2012 Austin P. Torney

All That Lies Between

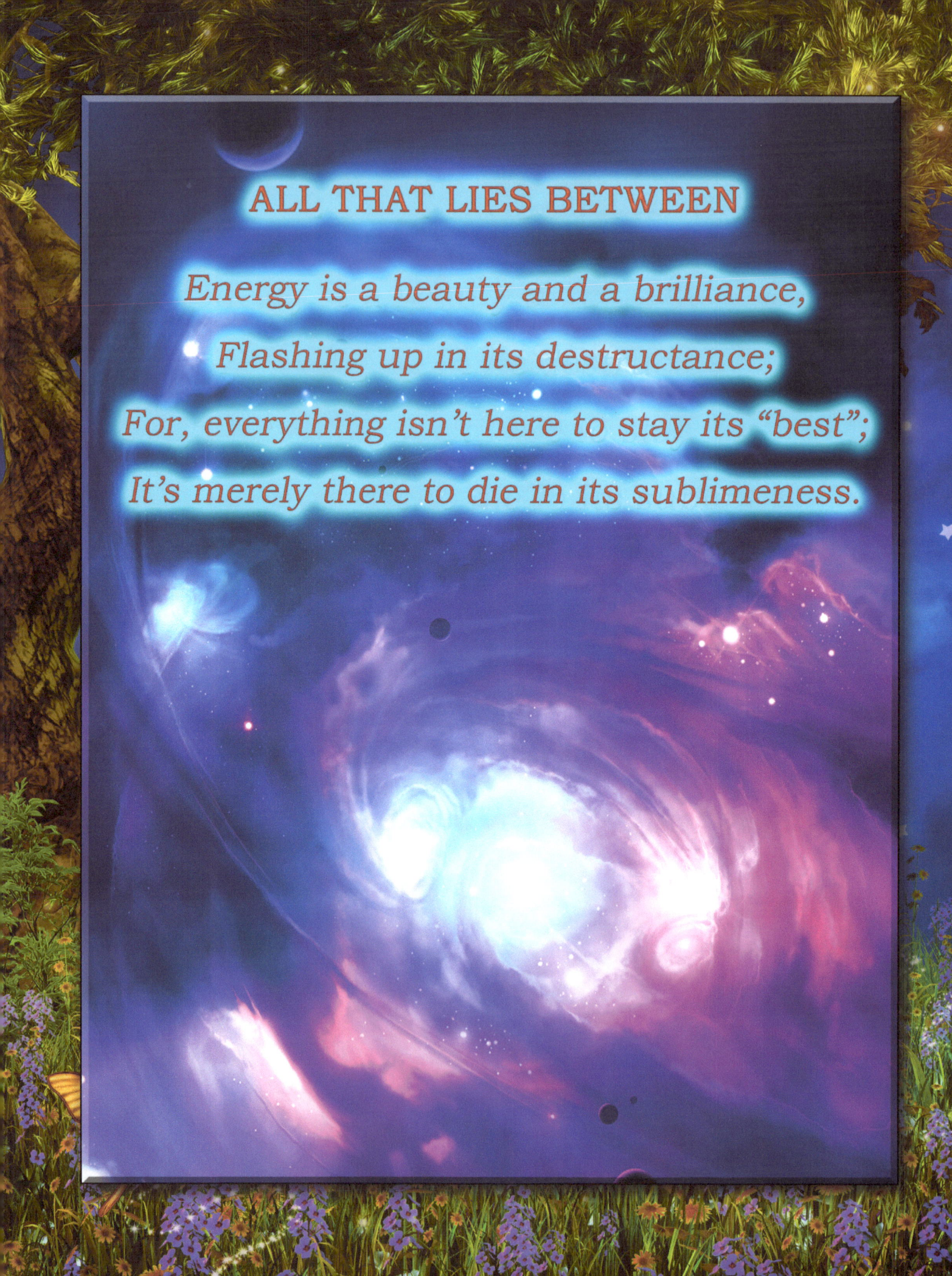

ALL THAT LIES BETWEEN

Energy is a beauty and a brilliance,
Flashing up in its destructance;
For, everything isn't here to stay its "best";
It's merely there to die in its sublimeness.

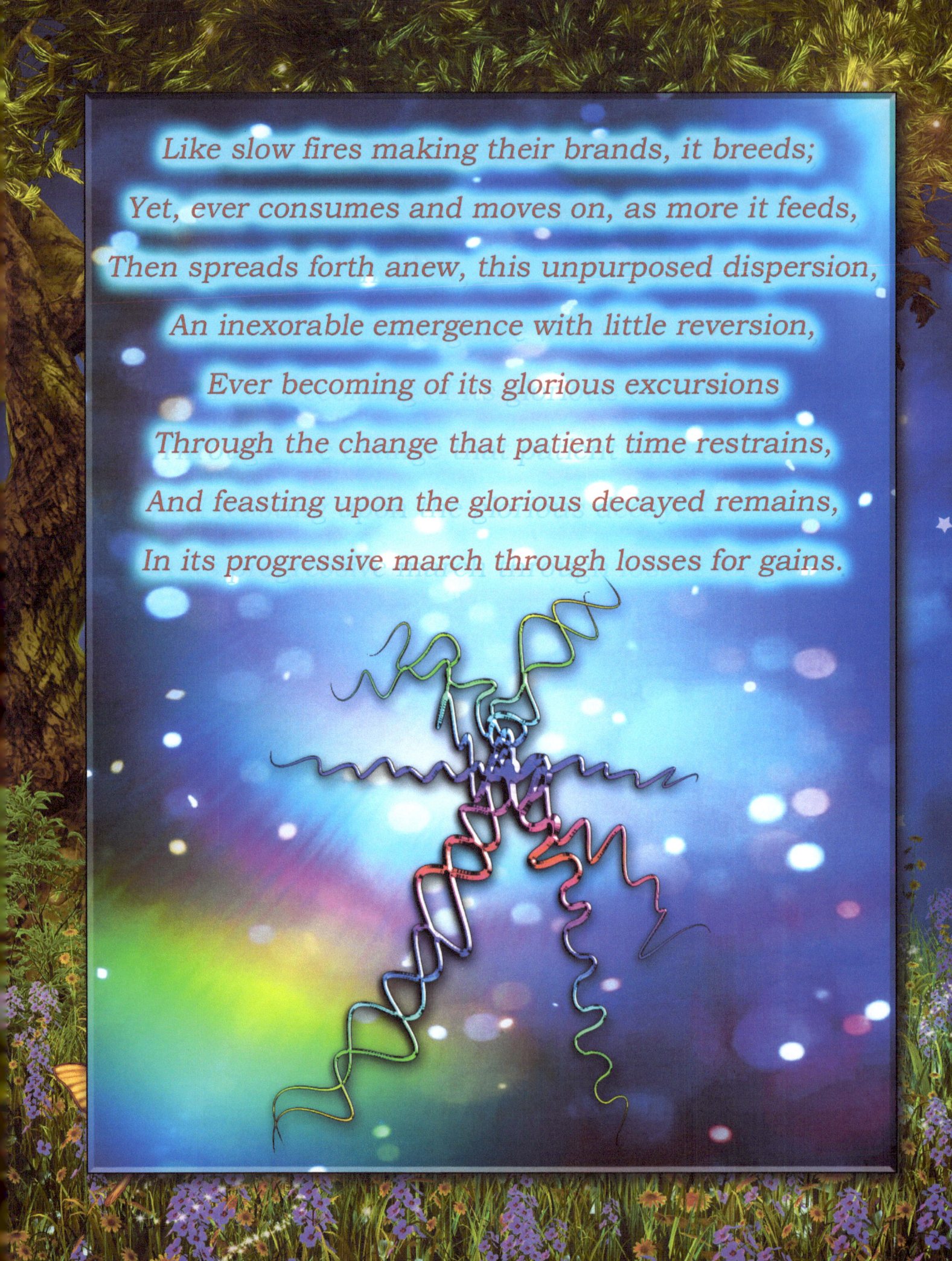

Like slow fires making their brands, it breeds;
Yet, ever consumes and moves on, as more it feeds,
Then spreads forth anew, this unpurposed dispersion,
An inexorable emergence with little reversion,
Ever becoming of its glorious excursions
Through the change that patient time restrains,
And feasting upon the glorious decayed remains,
In its progressive march through losses for gains.

We have oft described the causeless—
That which was always never the less,
As well as the beginnings of our quest,
And, too, have detailed, in the rarest of glimpses,
The slowing end of all of "forever's" chances.

So, then, we must now turn our attention keen
To all of the action that exists in-between—
All that's going on, and has gone before,
Out to the furthest reaches "ever-more";

For, everything that ever happens,
Including life and all our questions—
Meaning every single event ever gone on
Of both the animate and the non—
Is but from a single theme played upon.

This, then, is of the simplest analysis of all,
For it heeds mainly just one call—
That of the second law's dispersion,
The means for each and every occasion,
From the closest to the farthest range—
That which makes anything change.

These changes range from the simple,

Such as a bouncing ball resting still

Or ice melting that gives up its chill,

To the more complex, such as digestion,

Growth, death, and even reproduction.

There is excessively subtle change, as well,

Such as the formations of opinions tell

And the creation or rejections of the will.

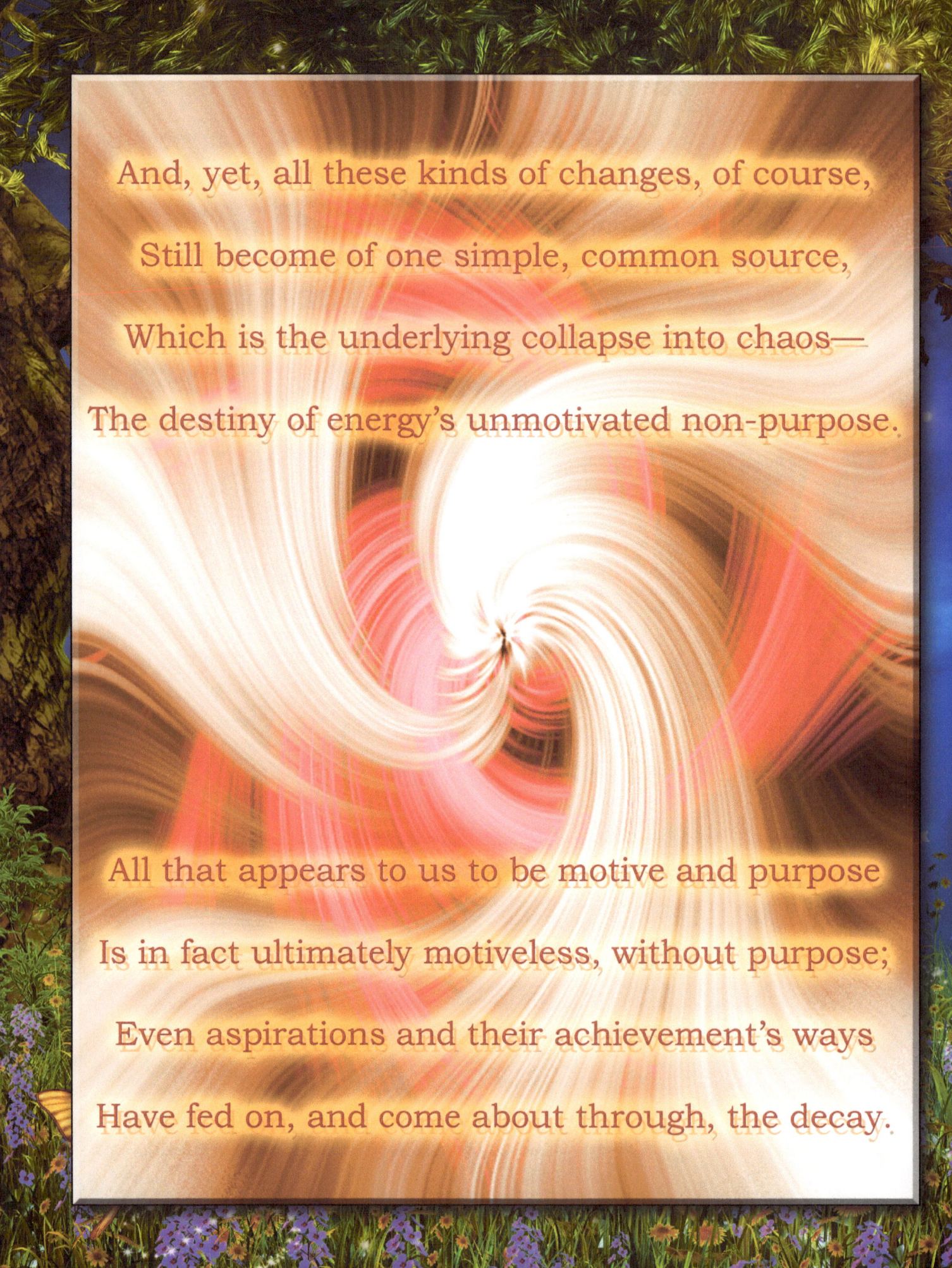

And, yet, all these kinds of changes, of course,

Still become of one simple, common source,

Which is the underlying collapse into chaos—

The destiny of energy's unmotivated non-purpose.

All that appears to us to be motive and purpose

Is in fact ultimately motiveless, without purpose;

Even aspirations and their achievement's ways

Have fed on, and come about through, the decay.

The deepest structure of change is but decay;

Although, it's not the quantity of energy's say

That causes decay, but the *quality,* for it strays.

Energy that is localized is potent to effect change,

And, in the course of causing change, it ranges,

Spreading, and becoming chaotically distributed,

Losing its *quality* but never of its quantity rid.

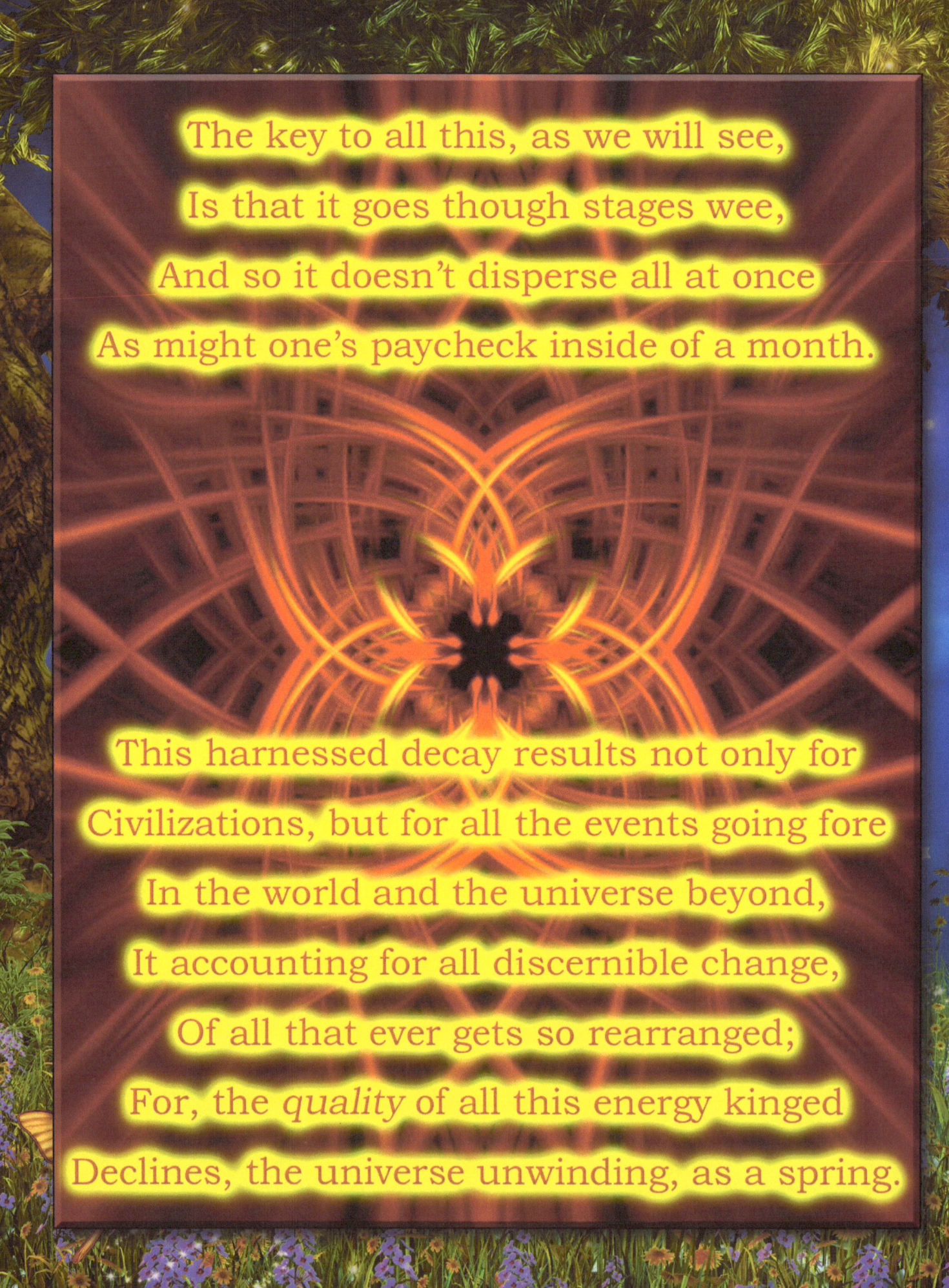

The key to all this, as we will see,
Is that it goes though stages wee,
And so it doesn't disperse all at once
As might one's paycheck inside of a month.

This harnessed decay results not only for
Civilizations, but for all the events going fore
In the world and the universe beyond,
It accounting for all discernible change,
Of all that ever gets so rearranged;
For, the *quality* of all this energy kinged
Declines, the universe unwinding, as a spring.

Chaos may temporarily recede,
Quality building up for a need,
As when cathedrals are built, or forms,
And when symphonies are performed;

But, these are but local deceits,
Born of our own conceits;
For, deeper in the world of kinds
The spring inescapably unwinds,
Driving its energy away—
As ALL is being driven by decay.

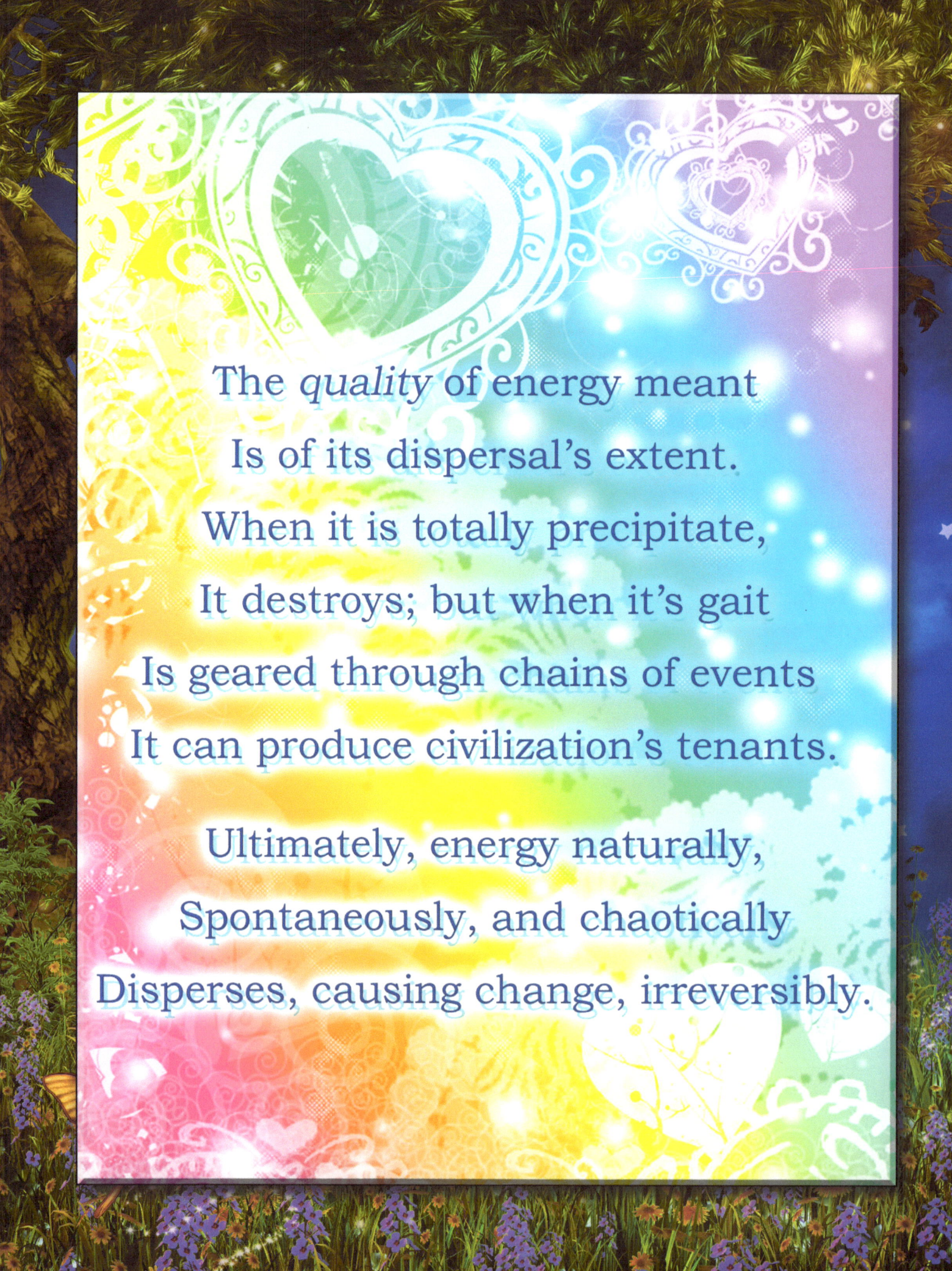

The *quality* of energy meant
Is of its dispersal's extent.
When it is totally precipitate,
It destroys; but when it's gait
Is geared through chains of events
It can produce civilization's tenants.

Ultimately, energy naturally,
Spontaneously, and chaotically
Disperses, causing change, irreversibly.

Think of a crowd of atoms jostling,
At first as a vigorous motion happening
In some corner of the atomic crowd—
They hand on their energy, loud,
Inducing close neighbors to jostle, too,

And soon the jostling disperses, too—
The irreversible change but the potion
Of the random, motiveless motions.

And such does hot metal cool, as atoms swirl,
There being so many atoms in the world

Outside it than in the block metal itself
That entropy's statisticals average themselves.

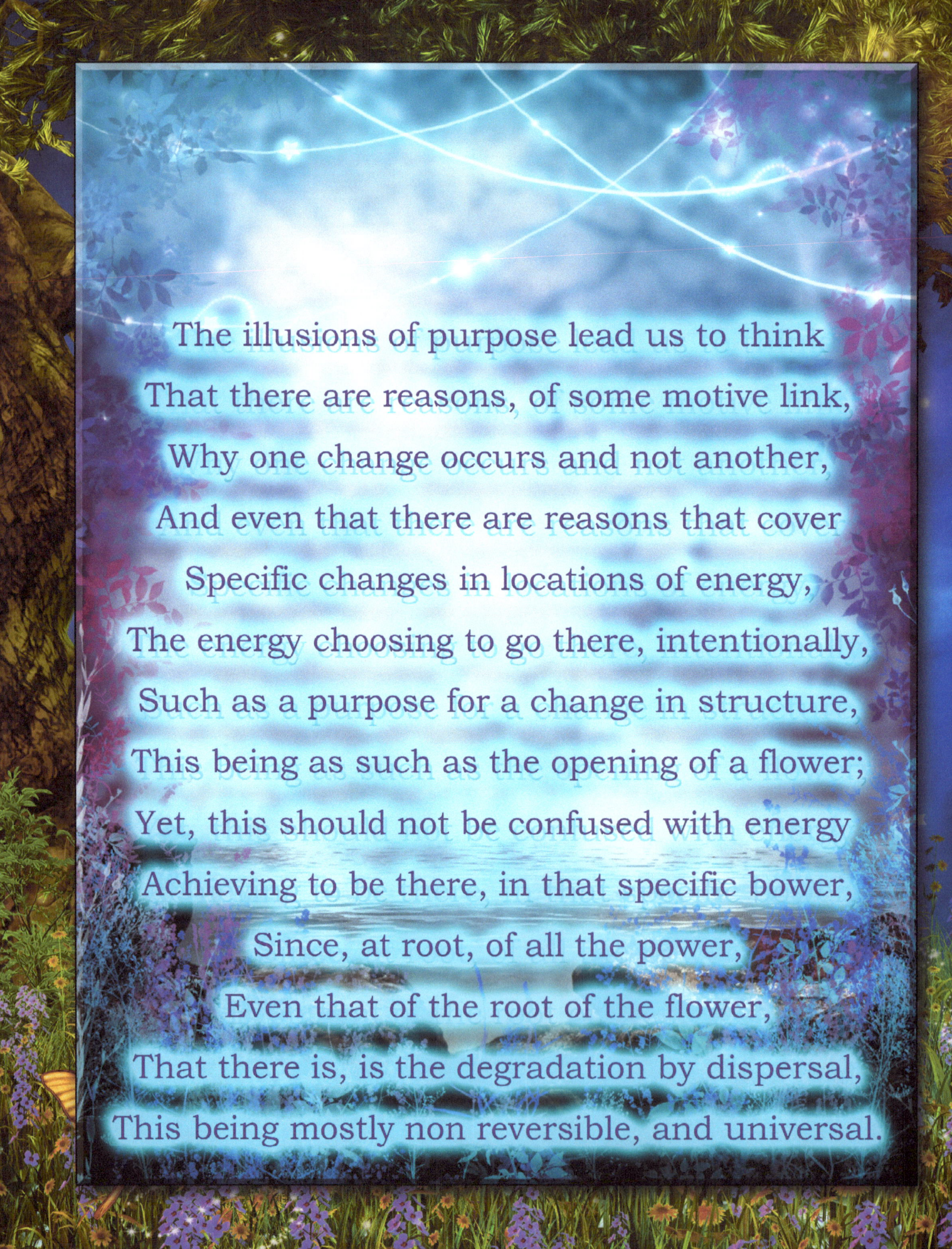

The illusions of purpose lead us to think
That there are reasons, of some motive link,
Why one change occurs and not another,
And even that there are reasons that cover
Specific changes in locations of energy,
The energy choosing to go there, intentionally,
Such as a purpose for a change in structure,
This being as such as the opening of a flower;
Yet, this should not be confused with energy
Achieving to be there, in that specific bower,
Since, at root, of all the power,
Even that of the root of the flower,
That there is, is the degradation by dispersal,
This being mostly non reversible, and universal.

The energy is always still spreading, thencely,

Even as some temporarily located density—

An illusion of specific change

In some region rearranged,

But, actually, it's just lingering there, "discovering",

Until new opportunities arise for exploring,

The consequences but of 'random' opportunity,

Beneath which, purpose still vanishes entirely.

Events are the manifestations

Of overriding probability's instantiations—

Of all of the events of nature, of every sod,

From the bouncing ball to conceptions of gods,

Of even free will, evolution, and all ambition;

For, they're of our simple idea's elaborations,

Although, for the latter stated there

And such for that as warfare,

Their intrinsic simplicity

Is buried more deeply.

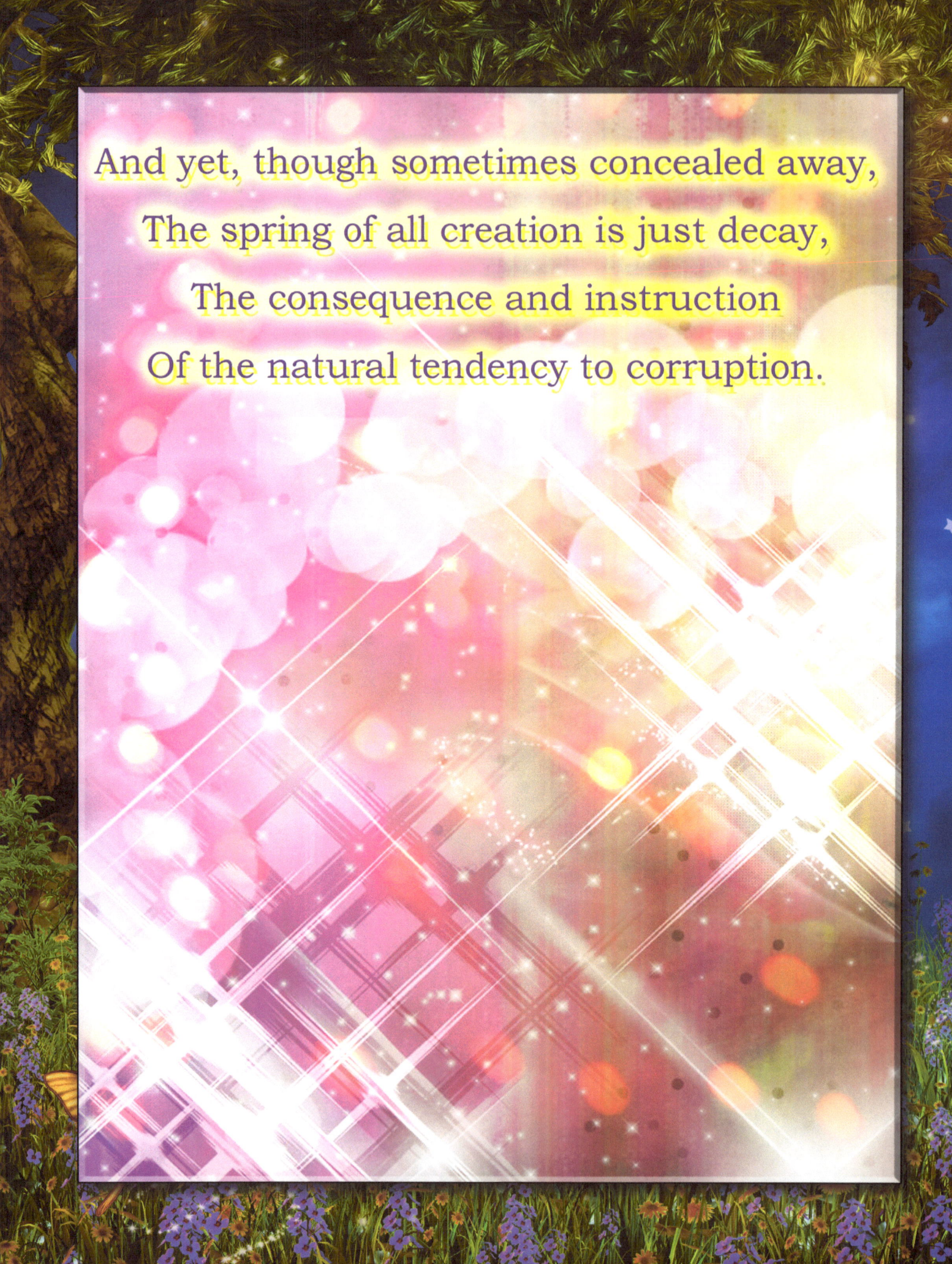

And yet, though sometimes concealed away,
The spring of all creation is just decay,
The consequence and instruction
Of the natural tendency to corruption.

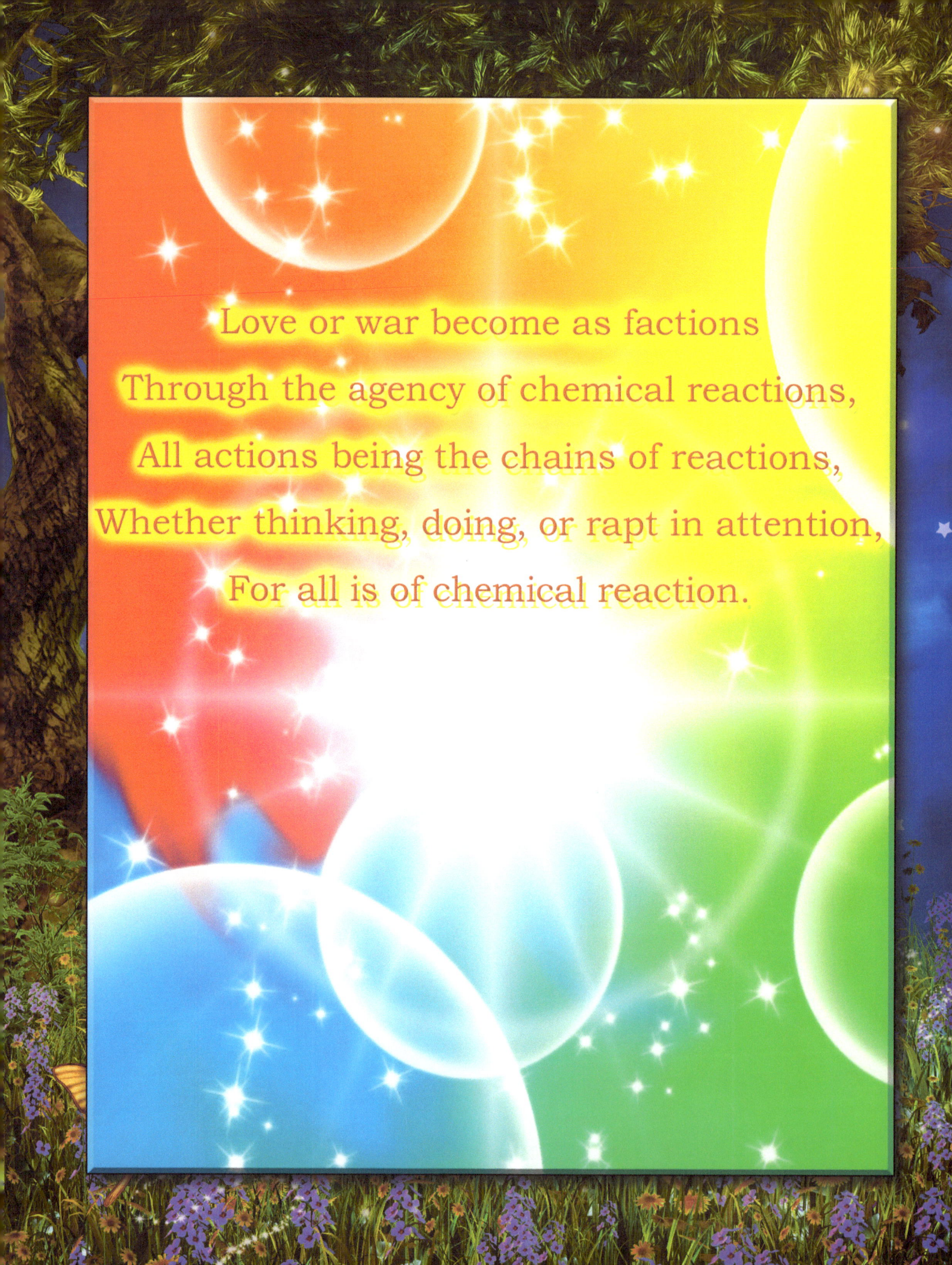

Love or war become as factions
Through the agency of chemical reactions,
All actions being the chains of reactions,
Whether thinking, doing, or rapt in attention,
For all is of chemical reaction.

At its most rudimentary bottom,

Chemical reactions are rearrangement of atoms,

These being species of molecules,

That, with perhaps additions and deletions

Then go on to constitute another one, by fate,

Although, they sometimes only change shape,

But, too, can be consumed and torn apart,

Either as a whole or in part; so cruel,

A source of atoms for another molecule.

Molecules have neither motive nor purpose to act—

Neither an inclination to go on to react

Nor any urge to remain unreacted;

So, then, why do reactions occur, if unacted?

Molecules are but loosely structured

And so they can be easily ruptured,

For reactions may occur if the process energy norm

Is degraded into a more dispersed and chaotic form,

And, so, as they usually are always constantly subject

To the tendency to lose energy as the abject

Jostling carries it away to the surroundations,

Reactions being misadventure's transformations,

It then being that some transient arrangements

May suddenly be "frozen" into "permanences"

As the energy leaps away to other experiences.

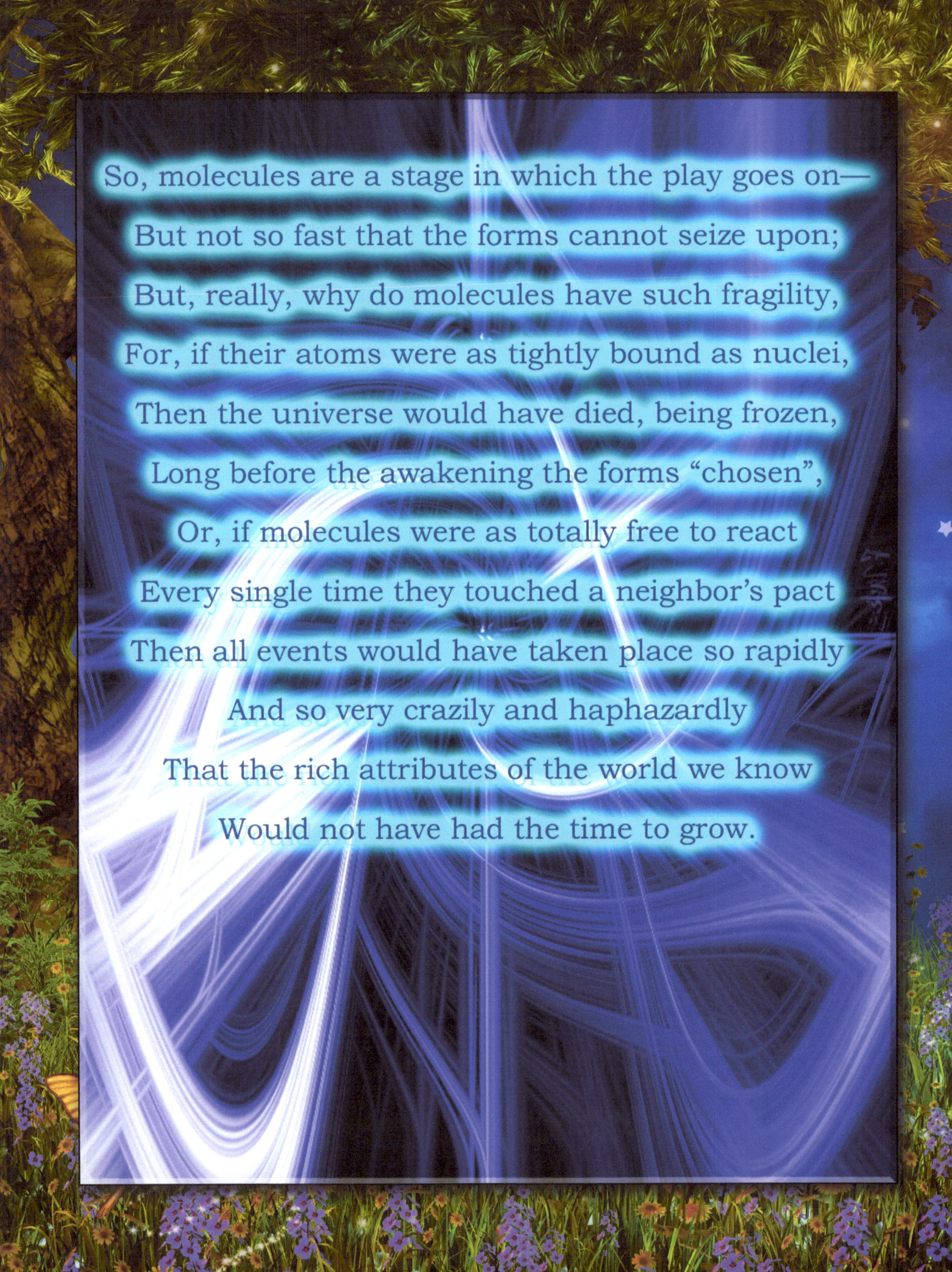

So, molecules are a stage in which the play goes on—
But not so fast that the forms cannot seize upon;
But, really, why do molecules have such fragility,
For, if their atoms were as tightly bound as nuclei,
Then the universe would have died, being frozen,
Long before the awakening the forms "chosen",
Or, if molecules were as totally free to react
Every single time they touched a neighbor's pact
Then all events would have taken place so rapidly
And so very crazily and haphazardly
That the rich attributes of the world we know
Would not have had the time to grow.

Ah, but it is all of the necessitated restraint,
For it ever takes time the scene to paint,
As such as in the unfolding of a leaf—
The endurations for any stepping feat,
As of the emergence of consciousness
And the paused ends of energy's restlessness:
Is of the controlled consequence of collapse
Rather than one that's wholly precipitous.

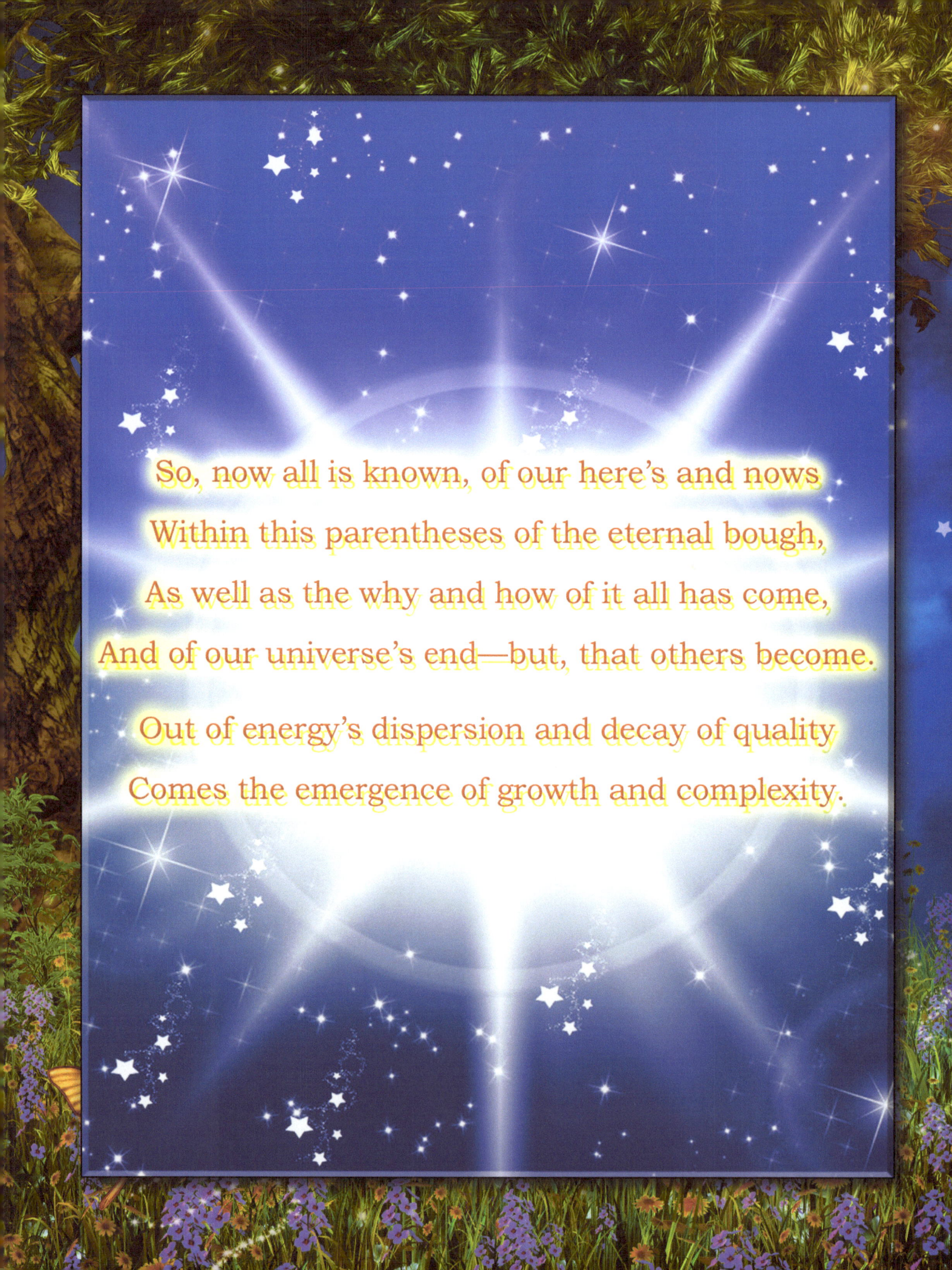

So, now all is known, of our here's and nows

Within this parentheses of the eternal bough,

As well as the why and how of it all has come,

And of our universe's end—but, that others become.

Out of energy's dispersion and decay of quality

Comes the emergence of growth and complexity.

(The verse lines,
being like
molecules, warmed,

Continually broke apart
and reformed

About the rhymes
which tried
to be nonintrusions,

Eventually all flexibly
stabilizing to conclusion.)